Grade K · Unit 3

Inspire Science

Weather and the Sun

Mc
Graw
Hill
Education

mheducation.com/prek-12

Send all inquiries to:
McGraw-Hill Education
8787 Orion Place
Columbus, OH 43240

ISBN: 978-0-07-699609-4
MHID: 0-07-699609-3

Printed in the United States of America.

9 10 11 LMN 26 25 24 23

Table of Contents
Weather and the Sun

Weather

The Sun and Earth's Surface

Weather

What do you wonder about the weather?

GO ONLINE

Check out
Stormy Weather.

Talk About It

Look at the photo.
Watch the video.
Describe what you see.
Draw a picture.

What Does a Meteorologist Do?

Meteorologists study weather.

They watch for patterns.

Meteorologists tell us what the weather will be each day.

GO ONLINE

Learn about meteorologists. Find out how science helps them do their jobs.

Look at the photo.
What does the girl need?
Color your answers.

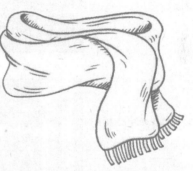

Weather

Here are some words you will learn.

clouds

drought

forecast

season

thermometer

thunderstorm

Words to Know

data   observe   pattern

Here are some words you will use in the lessons.

PAGE KEELEY
SCIENCE
PROBES

Thermometer

Sonia   Ava   Wendy

A thermometer is a weather tool.
What does a thermometer measure?
Who has the best idea?
Circle your answer.

Sonia: It measures the rain.

Ava: It measures the temperature.

Wendy: It measures how strong the
wind blows.

💬 **Talk About It**

Which friend do you agree with?
Tell a partner.

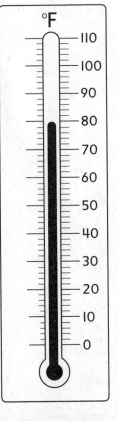

°F
110
100
90
80
70
60
50
40
30
20
10
0

LESSON 1

Describe Weather

What is happening in the woods?

▶ **GO ONLINE**

Check out *In the Woods*.

Draw a picture of your favorite kind of weather.

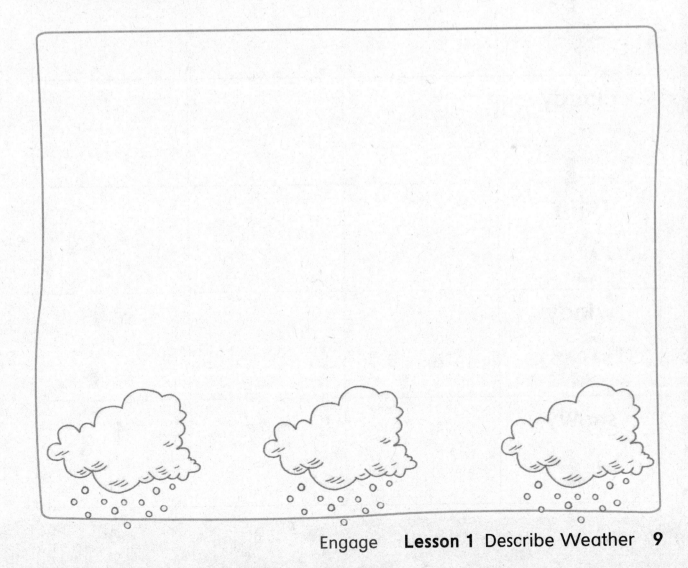

Data Analysis

Record the Weather

Make a Prediction How can we describe weather?

Investigate

Make a ✓ to record the weather for three days.

Weather	Day 1	Day 2	Day 3
sunny			
cloudy			
rainy			
windy			
snowy			

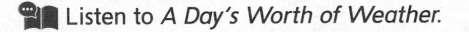 Listen to *A Day's Worth of Weather.*

1. (Circle) Ravi's weather on the picture and the table. Use a blue crayon.

2. (Circle) your weather in the table on page 10. Use an orange crayon.

Talk About It

Did you have the same kind of weather as Ravi? Show a partner.

Vocabulary

Look and listen for these words.

clouds

cool

rain

snow

temperature

thermometer

warm

weather

wind

Your Weather

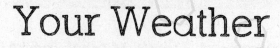 GO ONLINE

Watch the video *Measure and Describe Weather* to learn more about weather.

What is the weather today? Draw a picture.

Demonstration

Measure Weather

Make a Prediction How can we use tools to measure weather?

Investigate

You will use these tools to measure rain, wind, and temperature. Match the tool to what it measures.

temperature

wind

rain

Talk About It

What did you discover about tools? Show a partner.

Engineering

Make a Windsock

How can you tell which way the wind is blowing?

Build a Model

1. Make a windsock.

2. Measure the direction of the wind.

3. Improve your model.

4. Measure the direction of the wind.

Copyright © McGraw-Hill Education (1 5)Ken Cavanagh/McGraw-Hill Education,
(2)Stephen Ogilvy/McGraw-Hill Education, (3 4)Jacques Cornell/McGraw-Hill Education,
(6)McGraw-Hill Education

Materials

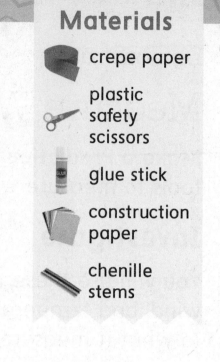

crepe paper

plastic safety scissors

glue stick

construction paper

chenille stems

5. Record Data How did the models
work? Draw a picture.

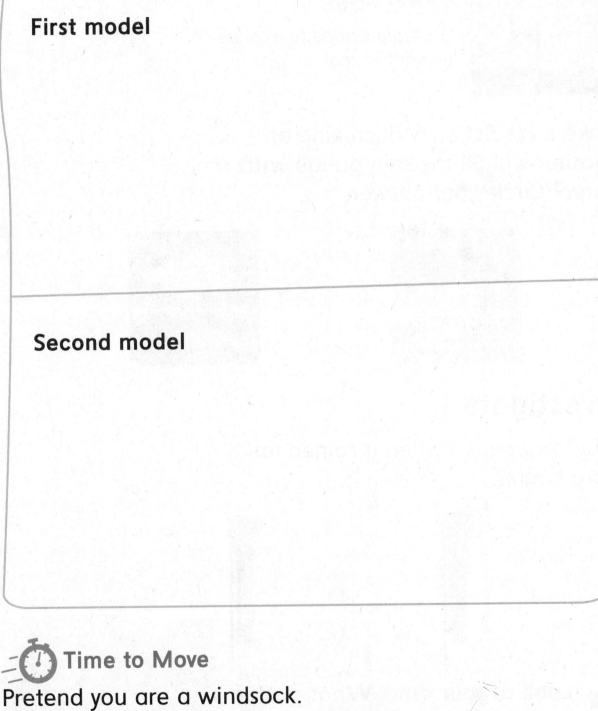

First model

Second model

⏱️ **Time to Move**

Pretend you are a windsock.
Move your arms and sway your body.

Simulation

Rain Gauge

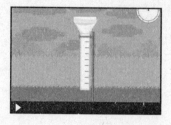

GO ONLINE

Use *Rain Gauge* to explore a weather tool.

Make a Prediction Which kind of weather will fill the rain gauge with water? Circle your answer.

Investigate

What happened when it rained for more hours?

 Look at your **data**. What **patterns** can you observe? Tell your class.

Hands On
Measure Rain

Make a Prediction How much will it rain? (Circle) your answer.

Materials

craft stick

plastic cup

marker

tape

some

a lot

Investigate

1. Use a cup and a craft stick to build a rain gauge.

2. Place your rain gauge outside.

3. Wait three days.

4. Observe.

5. **Record Data** Mark how much it rained. Color your craft stick.

MATH Connection

Where did it rain the most? Compare.

Review

EXPLAIN
THE PHENOMENON

| What is happening in the woods?

Draw a picture.

What is happening?

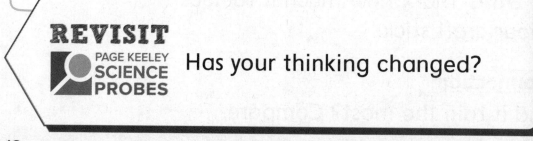

REVISIT
PAGE KEELEY
SCIENCE
PROBES

Has your thinking changed?

Module: Weather

LESSON 2 LAUNCH

Weather Patterns

Maria   Omar   Erin

Three friends are talking about weather. Who is describing a weather pattern?

Omar: It is very cold today.

Maria: January is colder than June.

Erin: It snowed for five hours.

Talk About It

Circle the picture of the friend you agree with. Show a partner.

LESSON 2

Weather Patterns

When do rainbows appear?

GO ONLINE

Check out *Rainbow.*

What are the colors in a rainbow?
Draw a picture. Use your crayons.

INQUIRY ACTIVITY

Copyright © McGraw-Hill Education (t to b)Jacques Cornell/McGraw-Hill Education, (2) McGraw-Hill Education, (3) Leena Robinson/Shutterstock.com

Hands On

Temperature

Make a Prediction How does temperature change during the day? Circle your answer.

It is warmer in the morning.

It is warmer in the afternoon.

Investigate

1. Measure the temperature in the morning.

2. Measure the temperature in the afternoon.

Materials

thermometer

crayons

3. Record Data Complete the table with your class.

Time of Day	Day 1	Day 2	Day 3
morning	cool warm	cool warm	cool warm
afternoon	cool warm	cool warm	cool warm

MATH Connection

How many cool mornings did you observe?

How many warm afternoons did you observe?

Count.

Vocabulary

Look and listen for these words.

fall   season   spring

summer   winter

Seasons

GO ONLINE

Watch the video *Patterns and Weather* to learn about the four seasons.

Listen to *Weather and Seasons.*

What is your favorite season?
What do you like to do?
What do you wear? Draw a picture.

Data Analysis

Compare Seasons

Make a Prediction What is the weather like in different seasons? Draw a line from the word to the photo.

Material

\\ crayons

spring

summer

fall

winter

⏱ **Time to Move**

Wave your arms when your teacher shares a photo of your favorite season.

Investigate

1. Look at the forecasts.

2. **Record Data** Complete the table with your class. Draw pictures.

Season	What I Wear	What I Do
spring		
summer		
fall		
winter		

 Talk About It

Do you need different clothes for each season? Show a partner.

Inspect

Read Look at the photo. Read the text.

Find Evidence

Reread Circle the clues in the photo. Underline the words that tell you about the temperature of the seasons.

Make Connections

💬 **Talk About It**

How did the photo help you know the season? How do people dress for each season?

Sledding Fun

A **season** is a time of the year.

Each year has four seasons.

Spring weather is warm.

Summer weather is hot.

Fall weather is cool.

Winter weather is cold.

Explain   **Lesson 2** Weather Patterns   **27**

Material

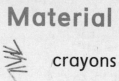 crayons

Data Analysis

Observe Clouds

Make a Prediction Are clouds always the same?

Investigate

1. Go outside.

2. Observe the clouds.

3. Observe the weather.

4. Record Data Circle the clouds you saw.

cumulus

cirrus

stratus

5. Record Data Draw the clouds.
Draw a symbol for the weather.

Day	Clouds	Weather
Day 1		
Day 2		
Day 3		
Day 4		
Day 5		

 Look at your **data.** How can clouds
tell us about **weather patterns**?
Tell a partner.

Simulation

Patterns and Seasons

GO ONLINE

Explore the seasons
in Sacramento and Baltimore.

Make a Prediction Which city has more
warm days in January? Circle your answer.

Baltimore	Sacramento

■ cold weather

▫ warm weather

Investigate

1. Choose a day in January.

2. Look at the pattern.

3. Predict the weather.

4. Repeat the steps for July.

MATH Connection

Look at the calendars.
Which city has more snowy days?

5. Record Data Draw pictures.

January

July

ENVIRONMENTAL Connection

How do weather changes affect people, plants, and animals?

Review

EXPLAIN
THE PHENOMENON | When do rainbows appear?

Draw a picture.

Weather before a Rainbow

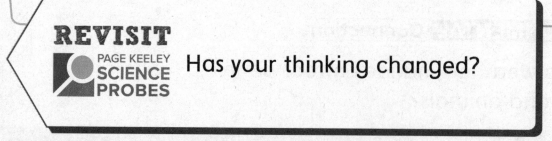

REVISIT
PAGE KEELEY
SCIENCE
PROBES

Has your thinking changed?

LESSON 3 LAUNCH

Forecast

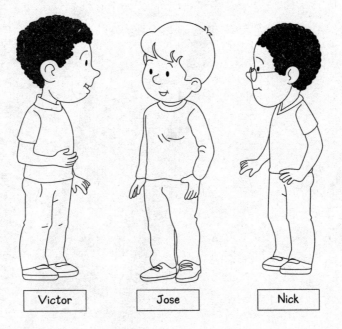

Victor   Jose   Nick

A weatherperson forecasts the weather.
What does it mean to forecast the
weather? Who has the best idea?

Victor: Describe what the weather will be like
next year.

Jose: Describe what the weather will be like
tomorrow.

Nick: Describe what the weather is like
right now.

💬 **Talk About It**

Talk about your thinking with
a partner.

°C

50

40

30

20

10

0

Forecast Weather

What do the thermometer symbols mean?

▶ **GO ONLINE**

Check out *Thermometer*.

Color the thermometer to match the weather you see in the photo. What is the weather like? Draw a picture.

INQUIRY ACTIVITY

Materials

crayons

forecast

Data Analysis

Tomorrow's Weather

Make a Prediction What will the weather be like tomorrow?

Investigate

1. Look at the forecast.

2. Observe the weather.

3. **Record Data** Draw a picture.

💬 **Talk About It**

Did the weather match your forecast? Tell a partner.

 Listen to *Storm Warning*.

1. Circle what Maisy's mom sees on the television.

2. Circle the weather.

 Talk About It

Did the weather match the forecast? Tell a partner.

Vocabulary

Look and listen for these words.

forecast

predict

GO ONLINE

Watch the video *Predict Weather* to learn more about how science and technology help forecast weather.

What are some other tools people use to predict weather?

PRIMARY SOURCE

The Farmer's Almanac was first printed in 1792.

Materials

crayons

forecast

Data Analysis

Forecast Weather

Make a Prediction Are forecasts always correct? Circle your answer.

Yes   **No**

Match each word to a weather symbol.

sunny   warm

rainy   cool

windy   snowy

💬 **Talk About It**

How can weather forecasts help us?
Tell a partner.

INQUIRY ACTIVITY

Investigate

1. Look at the forecast.

2. **Record Data** Draw a symbol for each day.

3. Draw a check mark if the forecast was correct.

MATH Connection

How many days was the forecast correct? Count your check marks.

Day	Forecast	Correct?
Day 1		
Day 2		
Day 3		
Day 4		
Day 5		

What questions do you have about the patterns in your data?

What Tools Do Meteorologists Use?

Meteorologists use many tools. These tools help them predict and forecast the weather.

They use thermometers and barometers. Thermometers measure temperature. Barometers measure air pressure.

Another tool meteorologists use is called an anemometer. This tool measures wind speed.

They rely on computers and satellites, too! It takes a lot of tools to forecast the weather.

Weather Tool

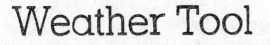

PRIMARY SOURCE

Four-cup anemometer, 1846

EXPLAIN
THE PHENOMENON | What do the thermometer symbols mean?

Draw a picture.

Weather Symbol	Weather

REVISIT

PAGE KEELEY
SCIENCE
PROBES

Has your thinking changed?

LESSON 4 LAUNCH

Severe Weather

Jin   Ling   Anna

How can you stay safe in a bad storm?

Jin: Go outside with coats and umbrellas.

Ling: Stay inside in front of a closed window.

Anna: Stay inside away from windows.

💬 Talk About It

Circle the friend you agree with.
Share your thinking.

LESSON 4

Severe
Weather

DISCOVER
THE PHENOMENON

What made these?

GO ONLINE

Check out *Hail*.

What was the weather like?
Draw a picture.

INQUIRY ACTIVITY

Copyright © McGraw-Hill Education. (t to b)Jacques Cornell/McGraw-Hill Education, (2)Sergiy Kuzmin/123RF, Ingram Publishing/SuperStock, (4)Jacques Cornell/McGraw-Hill Education, (5)McGraw-Hill Education, (6, 7)Jacques Cornell/McGraw-Hill Education, (b)Holly Curry/McGraw-Hill Education

Demonstration

Make Lightning

Make a Prediction How can you make lightning in a jar?

Investigate

1. Watch your teacher.

2. Rub the wool on the foam.

3. Touch the foam to the top of the jar.

4. Observe.

Materials

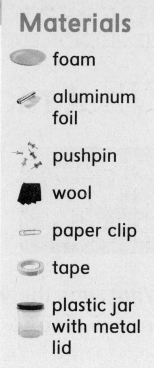

foam

aluminum foil

pushpin

wool

paper clip

tape

plastic jar with metal lid

5. Record Data What happened?
Draw a picture.

💬 **Talk About It**

Can you think of other ways you make lightning? Tell your teacher.

Vocabulary

Look and listen for these words.

blizzard

drought

hurricane

severe weather

thunderstorm

tornado

Severe Weather and You

GO ONLINE

Watch the video *Prepare for Severe Weather* to learn how to stay safe.

Listen to *Severe Weather*.

What kind of severe weather happens where you live? Draw a picture.

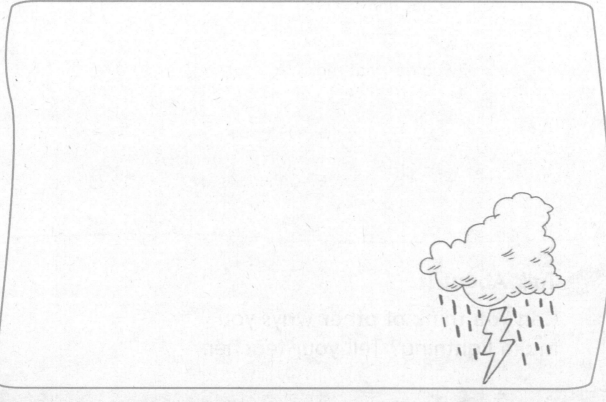

After the Storm

Hail comes in all shapes and sizes!

Draw the kind of weather that makes hail fall.

Demonstration

Rain, Rain, Go Away

Make a Prediction What happens when it rains for a long time? Circle your answer.

Investigate

1. Watch your teacher.

2. Record Data What happened?
Draw a picture.

Talk About It

Does it flood where you live?
Tell a partner.

ENVIRONMENTAL Connection

The photos on page 52 are from a
neighborhood in Texas before and after
Hurricane Harvey.

INQUIRY ACTIVITY

Copyright © McGraw-Hill Education (t to b)Janette Beckman/McGraw-Hill Education. Janette Beckman/McGraw-Hill Education, Ken Cavanagh/McGraw-Hill Education, James Turner/EyeEm/Getty Images, Yanavarut Phugongchana/Shutterstock

Hands On

Drought

Make a Prediction What happens when it does not rain for a long time? Circle your answer. Use a red crayon.

Materials

grass

water in measuring cup

aluminum pan

Investigate

1. Water some grass.

2. Do not water the other grass.

3. **Record Data** Look at your prediction. Circle the photo that looks most like the grass you did not water. Use a blue crayon.

4. Look at the photos.

5. Circle the photo that shows a drought.

ENVIRONMENTAL › Connection

How does drought affect people, plants, and animals?

Hands On

Make Thunder

Make a Prediction What materials can you use to make thunder? Circle your answers.

Materials

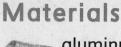 aluminum pan

 tin cans

 oatmeal container

 paper bag

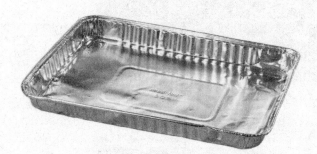

aluminum pan

tin cans

oatmeal container

paper bag

Investigate

1. Listen to the thunder.

2. Choose your materials.

3. Use your materials to make thunder.

 How can you use a **weather forecast** to help you **respond to severe weather**?

INQUIRY ACTIVITY

Prepare for Severe Weather

How can we stay safe during severe weather?

Materials

photo cards

plastic scissors

Investigate

1. Listen to the weather alert.

2. Look at the photo cards.

3. Find a match.

4. Watch your teacher.

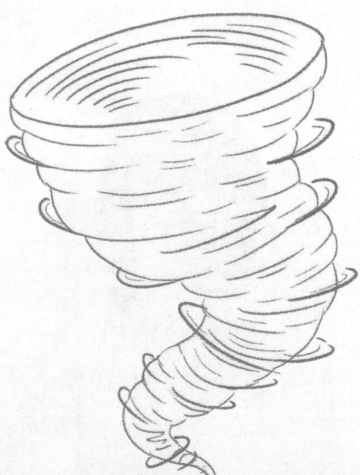

5. Record Data (Circle) the items that are useful in severe weather.

flashlight	batteries	basketball
canned food	nail polish	water
rubber duck	radio	guitar

Time to Move

How should you act during a tornado?
Show your teacher.

LESSON 4
Review

EXPLAIN
THE PHENOMENON

| What made these?

Draw a picture.

Kind of Weather

REVISIT
PAGE KEELEY
**SCIENCE
PROBES**

Has your thinking changed?

Use what you have learned to design your module project!

Make a Forecast

Be a meteorologist like Hugo.
Use words to describe the weather.

Build Your Model

1. Forecast the weather.

2. Organize the information.

3. Make a poster.

4. Choose your props.

5. Share the forecast.

Materials

crayons

magazines

poster board

plastic safety scissors

glue

Design Your Solution

Draw some ideas for your poster.

Module Wrap-Up

REDISCOVER
THE PHENOMENON

Tell what you learned. Draw a picture.

Look at your project if you need help.

The Sun and Earth's Surface

What is shining on the pinecone?

GO ONLINE

Check out *Sunlight in the Forest.*

💬 **Talk About It**

Look at the photo.

Watch the video.

Describe what you see.

Draw a picture.

What Does an Architectural Drafter Do?

Architectural drafters make drawings of buildings and other structures.

Architectural drafters use computers to make their drawings.

These drawings are called blueprints. They show what the building or structure will look like.

Blueprints show details. They give the measurements for each part.

► GO ONLINE

Learn about architectural drafters. Find out how architectural drafters use math and science to make blueprints.

Sam draws pictures of buildings.
Think about a place you like to visit.
Draw a picture of it.

What do you like about this place?

SAM
Architectural Drafter

The Sun and Earth's Surface

Here are some words you will learn.

Earth

protect

shade

shelter

Sun

surface

Words to Know

data   observe

Here are some words you will use in the lessons.

LESSON 1 LAUNCH

Warm Sand

Linda | Wei | Olivia

Why does the sand feel warm?
Who do you agree with?

Linda: *The wind warms the sand.*

Wei: *The Sun warms the sand.*

Olivia: *The water warms the sand.*

💬 **Talk About It**

Which friend do you agree with?
Tell a partner.

LESSON 1

Sunlight on Earth's Surface

How will the sunlight change the water?

GO ONLINE

Check out *Misty Morning.*

Draw a picture that shows one way you get hot on a sunny day.

INQUIRY ACTIVITY

Sunlight and Water

Make a Prediction What will sunlight do to water? Circle your answer.

make water cool

make water warm

Investigate

1. Put a thermometer in each cup of water.

2. **Record Data** Show how warm the water is in each cup. Color the thermometers.

3. Place one cup in sunlight.

4. Place the other cup where there is no sunlight.

5. Wait.

6. Show how the sunlight changed the water. Color the thermometers for each cup.

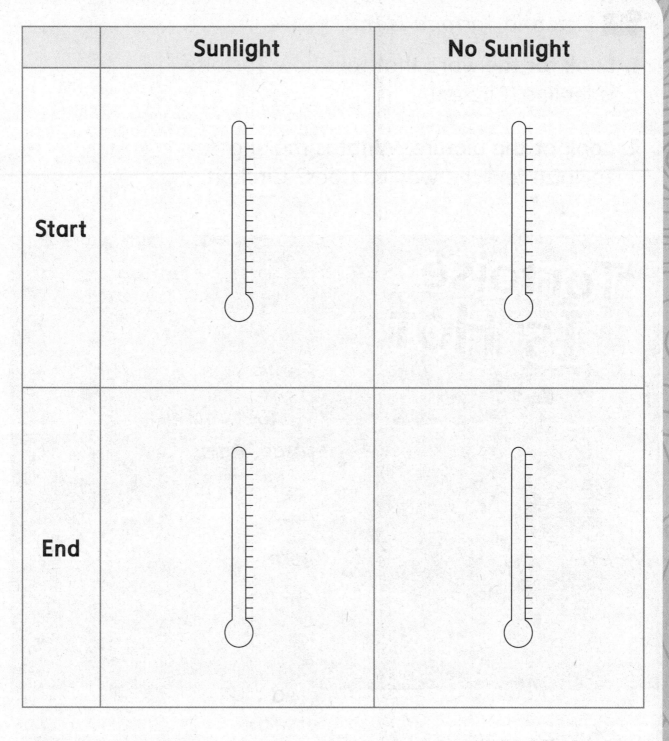

	Sunlight	**No Sunlight**
Start		
End		

Talk About It

What happened to the water?
Show a partner.

Sunlight in the Desert

💬📖 Listen to *Tortoise Is Hot*.

1. Look for the word that tells how Tortoise is feeling. Circle it.

2. Look at the picture. What is making Tortoise feel the way he does? Circle it.

⏱️ **Time to Move**

How can you cool off when it is hot? Show a friend.

The Sun

▶ **GO ONLINE**

Watch the video *The Sun Warms Earth.* Learn how the Sun changes Earth's surface.

💬📖 Listen to *Earth and the Sun.*

Why is Earth's surface warmer during the day than at night?

Draw a picture.

Show where the Sun is in the sky.

Vocabulary

Look and listen for these words.

Earth

Sun

surface

Materials

rocks

sand

soil

2 plates

plastic gloves

safety goggles

Hands On

Surfaces and Sunlight

Make a Prediction What happens to rocks, soil, and sand in the Sun? Circle your answer.

They become warm.

They become cool.

Investigate

BE CAREFUL Wear safety goggles and gloves.

1. Put the rocks, soil, and sand on one plate.

2. Place the plate in the Sun.

3. Put rocks, soil, and sand on another plate.

4. Put the plate where there is no sunlight.

5. Wait. Touch the rocks, soil, and sand on both plates.

6. Record Data Circle the word that tells about the rocks, sand, and soil.

Surface	Sunlight		No Sunlight	
rocks	warm	cool	warm	cool
sand	warm	cool	warm	cool
soil	warm	cool	warm	cool

 What was the **effect** of **sunlight** on the rocks, soil, and sand?

Demonstration

Melt in the Sunlight

Make a Prediction Which things will melt in the Sun? Circle your answers. Use an orange crayon.

crayons

eraser

pencil

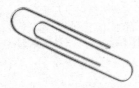

paper clip

chocolate

butter

ice cream

ice

cheese

Investigate

1. Observe.

2. **Record Data** (Circle) the things that melted. Use a blue crayon.

MATH Connection

How many things did the sunlight melt? Count.

 Talk About It

Did your prediction match the results? Show a partner.

Review

EXPLAIN
THE PHENOMENON | How will the sunlight change the water?

Draw a picture.

Sunlight's Effect on Water

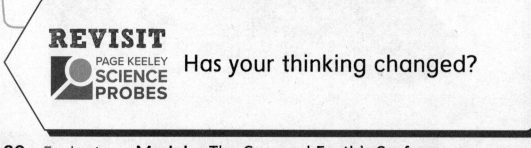

REVISIT
PAGE KEELEY
SCIENCE
PROBES
Has your thinking changed?

Sunlight and Shade

Kenny and Jim are playing outside.
It is hot and sunny.

Which boy has the best idea?

Kenny: *Let's play in the shade to stay cool.*

Jim: *The shade is darker, not cooler.*

💬 Talk About It

Talk about your thinking
with a partner.

LESSON 2

Protection from the Sun

DISCOVER
THE PHENOMENON

Why are the girls inside the tent?

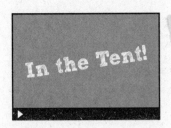

 **GO ONLINE**

Check out *In the Tent!*

Where is one place you go to get out of the sunlight? Draw a picture.

INQUIRY ACTIVITY

Material

red crayon

Simulation

Temperatures Throughout the Day

▶ GO ONLINE

Explore *Temperatures Throughout the Day* to learn about Sun, shade, and temperature.

Make a Prediction How does shade change temperature? Circle your answer.

Temperature goes up in shade.

Temperature goes down in shade.

Investigate

1. Choose the time of day.

2. Click on the middle Sun.

3. Click on the umbrella.

4. Observe what happens to the temperature.

5. Record Data Fill in the thermometers.

No Shade	Shade
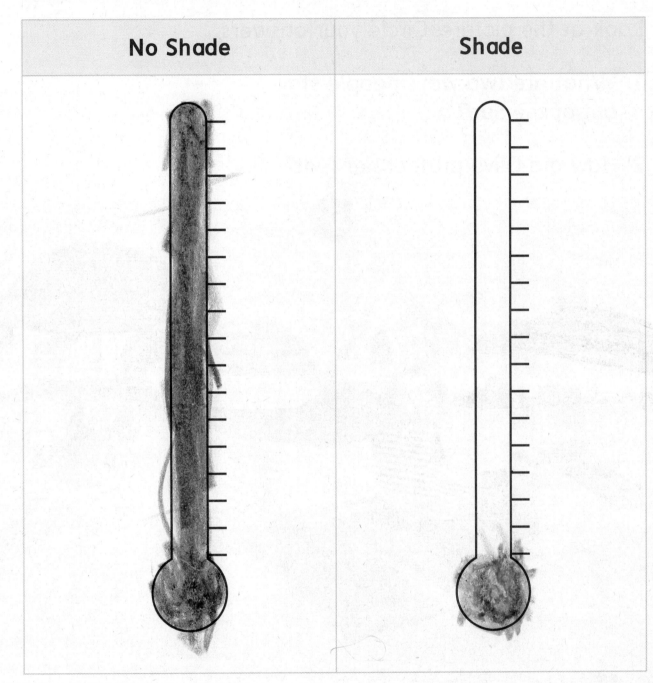

💬 **Talk About It**

What caused the temperature to change? Tell your class.

Stay Out of the Sunlight

💬📖 Listen to *A Day at the Beach*.

Look at the picture. Circle your answers.

1. What are two ways people stay out of the Sun?

2. How did Olive protect her feet?

⏱️ **Time to Move**

Show a friend how you walk on hot sand.

Shade

GO ONLINE

Watch the video *Shade from the Sun*. Learn how shade can help you stay cool on a hot, sunny day.

📖 Listen to *Made in the Shade*.

Where is one place people can find shade on a sunny day?

Draw a picture.

Vocabulary

Look and listen for these words.

protect

shade

shelter

Simulation

Temperatures Throughout the Day

▶️ **GO ONLINE**

Explore *Temperatures Throughout the Day* to learn more about shade and temperature.

Make a Prediction Which structure will make the most shade? Circle your answer.

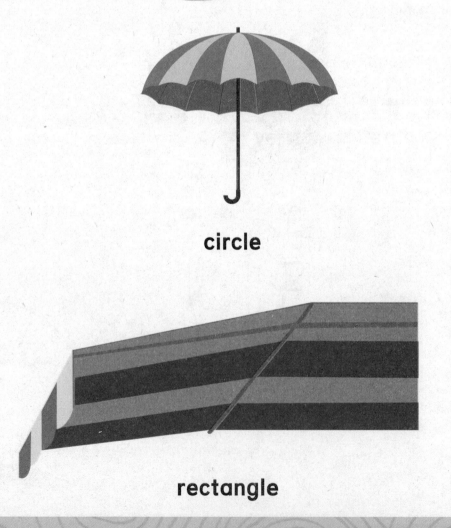

circle

rectangle

Investigate

1. Click on the middle Sun.

2. Click on the umbrella.
   Observe the shade.

3. Click on the canopy.
   Observe the shade.

MATH Connection What shape is the canopy? Where is the canopy in relation to the shade? Use position words.

💬 **Talk About It**

Compare the umbrella and the canopy. Tell a partner how they work.

INQUIRY ACTIVITY

Hands On

Sunscreen and Protection

Make a Prediction What will sunscreen do?

Investigate

1. Put sunscreen on plastic wrap.

2. Place plastic wrap over half of the paper.

3. Put the paper in a sunny area.

4. Wait.

5. Record Data Show what sunscreen does.
Draw a picture.

With Sunscreen	Without Sunscreen

 What was the **effect** of **sunlight** on the paper?

Inspect

How can you stay safe from the Sun? Look at the poster.

Find Evidence

Underline the title of the poster.

Make Connections

 Talk About It

The author says that you should wear clothes that protect you. Which clothes in this poster offer some protection? Show a partner.

Sun Wise with Shade Foundation

Avoid sun tanning beds

Apply sunscreen spf 30+

SUN SPF 30 + Block

Wear protective clothing

Seek shade!

Watch UV index forecast

UV 3

M T W TH F
90 90 90 90 90

Be cautious near water and sand

Be Sun Wise!

Adorabella
Grade 2
Brooklyn, New York

What Does a Civil Engineer Do?

Lili Hernandez is a civil engineer.

She designs structures that keep us safe and improve our lives.

Do you want to be a civil engineer like Lili?

She says you should play with blocks, build models, and work on puzzles!

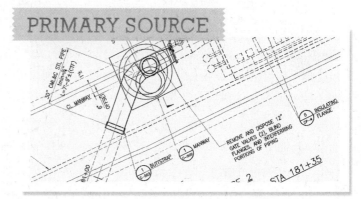

Review

EXPLAIN
THE PHENOMENON

Why are the girls inside the tent?

Draw a picture.

Outside Tent	Inside Tent

REVISIT

PAGE KEELEY SCIENCE PROBES

Has your thinking changed?

Use what you have learned to design your module project!

Design a Structure to Make Shade

Be an architectural drafter like Sam. It is a sunny day. You want to read outside. How can you build something to make shade?

Build Your Model

1. Design your structure.

2. Build your design.

3. Go outside. Test your design.

4. Make your design better.

Materials

butcher paper

tape

plastic safety scissors

cardboard box

cardboard tube

UV beads

Design Your Solution

How did you improve your design?
Draw a picture.

 Talk About It

What caused your design
to change?

Module Wrap-Up

REDISCOVER
THE PHENOMENON
Tell what you learned. Draw a picture.

Look at your project if you need help.

A

air the invisible gas that people and animals breathe

animal a living thing that moves and eats food

arctic very cold, freezing

B

blizzard a strong snowstorm that lasts a long time

burrow a hole or tunnel in the ground that an animal makes to live in

C

clouds groups of water droplets or ice in the sky

collide to hit something with strong force

cool somewhat cold; not warm or hot

D

dam a wall built across a river to stop the flow of water

desert a large area of land that gets very little rain

direction the course or path on which something moves

distance the amount of space between two places or things

drought a long period of time with little or no rain

Shutterstock.com, Matyas Rehak Shutterstock.com, Fkajfanc/iStock/Getty Images, mjrahi Publishing/age fotostock, Joggie Botma/Alamy Stock Photo, Photo by Tim McCabe, USDA Natural Resources Conservation Service

E

 Earth the planet on which we live

 **ecosystem** a group of living and nonliving things in an environment

 environment all the living and nonliving things in an area

F

 **fall** the season after summer

 farm a large area of land used for growing crops or raising animals

 **force** a push or pull

 **forecast** to say that something will happen by using information about the weather

forest a place where there are many trees

G

garden an area of land used for growing flowers or vegetables

H

habitat a place where plants and animals live

human a person

hurricane a strong storm with heavy rain and winds that blow in a circle

L

lake a body of water surrounded by land

light something that lets you see

living a thing that grows, changes, and has needs

M

materials what objects are made of

meadow a field of grassy land

motion a change in an object's position

N

natural resource something that comes from Earth that people use

need something you must have in order to live

nonliving a thing that does not grow, change, or have needs

nutrient something that living things need to grow

O

object anything that can be seen and touched

P

plant a living thing that has leaves and roots and makes its own food

pond a small body of fresh water

position the place where something is located

predict to say what will happen in the future

protect to keep safe from harm

pull a force that moves something closer to you

push a force that moves something away from you

R

rain water droplets that fall from clouds to Earth

recycle to make something new from something old

 **reduce** to use less of something

 reuse to use something again

 river a large natural stream of water

S

 **season** a time of year

 **severe weather** dangerous weather conditions

 shade the dark area caused when light is blocked

 **shelter** a place that gives protection from weather or danger

 **snow** soft, white crystals of ice that fall to earth

 **speed** how fast or slow something moves

 spring the season after winter

 summer the season after spring

 **Sun** the star closest to Earth

 **surface** the outside of something

 **survive** to live and grow

T

 temperature how hot or cold something is

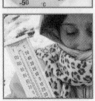 **thermometer** a tool that measures temperature

thunderstorm a storm with thunder, lightning, rain and wind

tornado a strong storm with winds that form a cloud that looks like a funnel

W

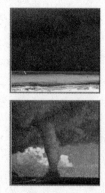

warm somewhat hot; not cool or cold

water the liquid that falls as rain from the sky

weather what the sky and air are like each day

wind air that is moving over the earth

winter the season after fall

✂ cut on all dashed lines   ⬜ fold on all solid lines

The **weather** is what the sky and air are like each day.

Memory Maker: Draw a weather forecast showing weather patterns.

forecast

Module: Weather   **VKV1**

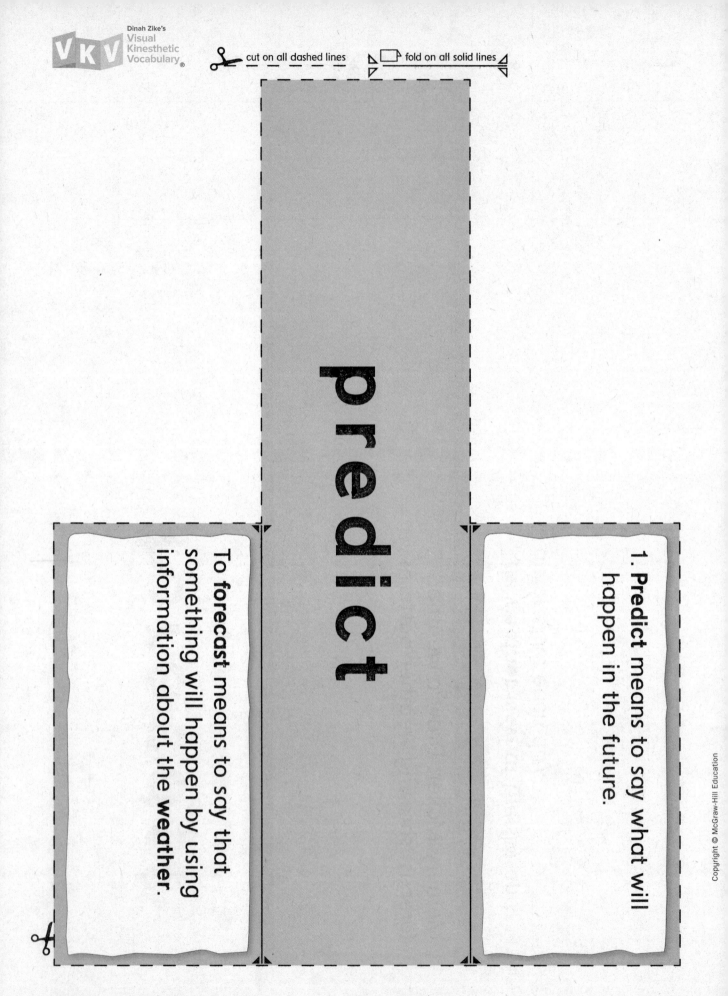

VKV® Dinah Zike's Visual Kinesthetic Vocabulary®

✂ cut on all dashed lines

fold on all solid lines

predict

To **forecast** means to say that something will happen by using information about the **weather**.

1. **Predict** means to say what will happen in the future.

✂ cut on all dashed lines

fold on all solid lines

VKV2   Module: Weather

Shelter is a place that gives protection from weather or danger.

Earth is the planet on which we live.

When an object blocks light from the **Sun**, it makes **shade.**

Shade is the dark area caused when light is blocked.

Earth

The **Sun** is the star closest to **Earth.**

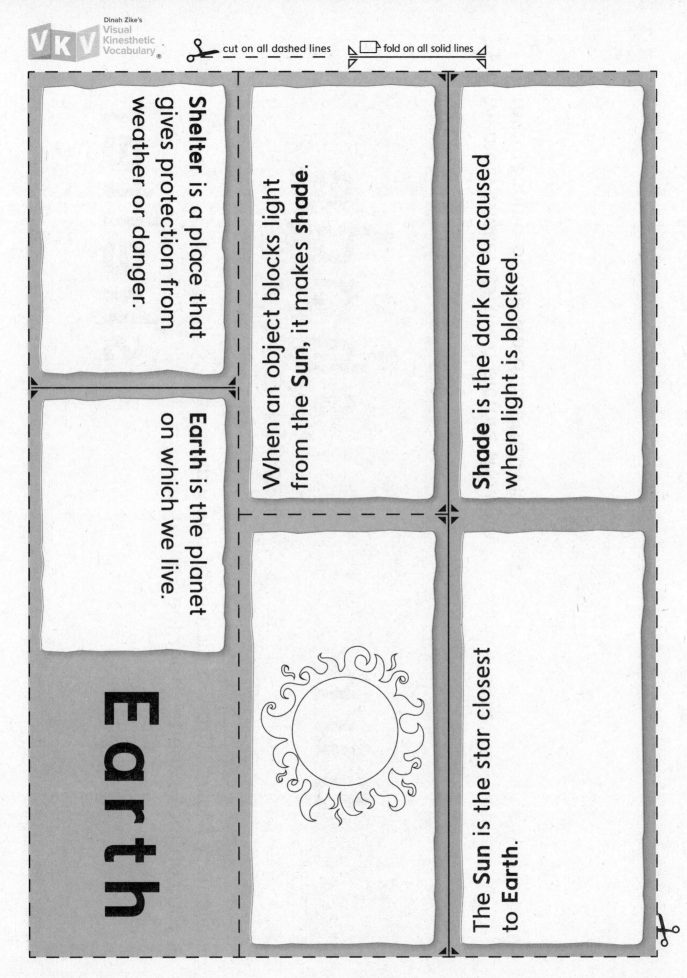

Module: The Sun and Earth's Surface   **VKV3**

Memory Maker: Draw a picture showing the meaning of the words on this VKV.

shade

Sun

shelter

Memory Maker: Draw a picture of one way you can make shelter from the Sun.

VKV4   Module: The Sun and Earth's Surface